Lessons Learned
from the Deepwater Horizon Response

National Institute for Occupational Safety and Health

December, 2011

Department of Health and Human Services
Center for Disease Control and Prevention
National Institute for Occupational Safety and Health

Cover photos by Rob Wolfe and Aaron Sussell, courtesy of NIOSH

Foreword

The explosion on the Deepwater Horizon disaster oil rig on April 20, 2010 resulted in the death of 11 workers and injury to another 17 workers. In the weeks and months after, as large amounts of crude oil released from the Macondo Well, tens of thousands of workers engaged in on- and off-shore containment and clean-up activities. Addressing concerns about the potential effects of the spill on human and environmental health in the Gulf, including potential risk to response workers, prompted an unprecedented response from agencies all across the Federal government.

On May 3, 2010, experts from the National Institute for Occupational Safety and Health (NIOSH) arrived on site at the Gulf of Mexico. The role NIOSH had in this interagency effort was to anticipate and address the occupational safety and health needs of the containment and cleanup response workers in close collaboration with the Occupational Safety and Health Administration (OSHA). As part of these activities, NIOSH led efforts in rostering workers, conducting health hazard evaluations, providing technical guidance by means of a joint OSHA/NIOSH publication, conducting health surveillance activities, and performing toxicity testing.

The Deepwater Horizon Response (DWHR) effort presented unique challenges in protecting response workers spread across the Gulf region, who performed a wide range of activities in physically and emotionally demanding circumstances. The DWHR presented opportunities to further expand our knowledge and understanding of protecting workers in complex, large-scale emergency responses. With the publication of "Lessons Learned from the Deepwater Horizon Response," NIOSH hopes to share the knowledge gained during this response (including the application of knowledge gained from past large-scale emergency responses), report how we can improve our response to similar events in the future, and facilitate a dialogue between NIOSH and partners in the government, industry, labor and academia on ways to improve the overall response to both natural and man-made disasters.

As we build our knowledge of what happens during the response to disasters, so too will we increase our understanding of how to protect response workers.

John Howard, MD
Director, National Institute for Occupational Safety and Health

Table of Contents

Executive Summary 1

NIOSH Response Activities 6
    NIOSH Rostering Effort 6
    Conducting Health Hazard Evaluations (HHEs) 8
    NIOSH Technical Guidance and Communications 8
    Health Surveillance 9
    NIOSH Toxicity Testing Related to the Deepwater Horizon Response 10

Lessons-Learned Observations 11
    General Deployment Issues 11
    Rostering Activities 12
    Conducting Health Hazard Evaluations (HHEs) 13
    NIOSH Technical Guidance and Communications 14
    Health Surveillance Activities... 14
    Other Health Assessment Efforts 15
        Biological Monitoring 15
        Long-Term Studies 15
        Mental Health 16

Follow-up Actions and Next Steps 17

Acknowledgements 18

Executive Summary

Following the unprecedented federal response to the Deepwater Horizon (DWH) disaster, the National Institute for Occupational Safety and Health (NIOSH) conducted a review of the Institute's response activities. The purpose of this report is to evaluate those response activities, and in light of knowledge gained, identify ways to improve our response in the future. This report is organized by first providing information on NIOSH response activities, followed by several sections containing lessons-learned information, and ending in a discussion about next steps.

Occupational safety and health experts from NIOSH arrived on site in response to the oil spill on May 3, 2010, at the invitation of the Occupational Safety and Health Administration (OSHA), as part of the federal interagency effort to anticipate and address the occupational safety and health needs of containment and cleanup response workers on the Gulf Coast. The unprecedented circumstances and magnitude of the disaster posed numerous challenges for NIOSH.

NIOSH staff deployed to multiple locations in the five-state Gulf region; coordinated cross-agency and interagency operations in Atlanta, GA, and Washington, DC; supported other agencies' missions as subject matter experts; and provided technical and administrative support from NIOSH divisions, labs, and offices located throughout the country. This effort was a large activation of NIOSH personnel, cumulatively resulting in deployment of 106 staff into the field and involving close to 250 staff in total.

NIOSH successfully and rapidly performed multiple activities to protect worker safety and health in the Gulf, as highlighted below:

- For the first time during an event, developed a voluntary roster of workers to obtain a record of workers who participated in the containment and cleanup, creating a mechanism for locating and contacting them about possible work-related symptoms of illness or injury;

- Rapidly conducted health hazard evaluations associated with reported illnesses among workers involved in the Gulf response. These health hazard evaluation activities included assessments of complex exposures to heat, physical stress, fatigue, psychosocial and work organization factors, and toxic chemical and physical agents in numerous work tasks on the water and on the shore;

- Partnered with the Occupational Safety and Health Administration (OSHA) in the Department of Labor to develop interim guidance for Protecting Deepwater Horizon response Workers and Volunteers;

- Worked with OSHA, the U.S. Department of Health and Human Services (HHS) Assistant Secretary of Preparedness and Response (HHS/ASPR), the Substance Abuse Mental Health Services Administration (HHS/SAMHSA), the National Institute of Environmental Health Sciences (HHS/NIH/NIEHS), and the U.S. Coast Guard to provide the Unified Area Command (UAC) and other federal and state partners, BP, and workers guidance and communication/educational materials for protecting response workers;

- Conducted health surveillance by analyzing injury and illness data to increase understanding and awareness of the risks associated with Gulf oil response work;

- Designed, gathered materials and equipment, and began laboratory animal acute toxicity studies on the dispersant, crude oil, and dispersant/crude oil mixtures.

While NIOSH views its participation in the oil spill response as highly successful, a review and assessment of response activities and outcomes can identify gaps, issues, and problems that can point to revisions in processes and procedures that can improve future responses. Lessons-learned in response to this disaster, as outlined below, included those related to general deployment issues, rostering activities, health hazard evaluations, technical guidance and communications, health surveillance activities, toxicity testing, and other health assessment efforts.

General Deployment Issues

NIOSH responded to the Gulf initially at the invitation of OSHA. NIOSH received independent authority to provide on-site assistance once an agreement was signed between the Federal On-Scene Coordinator and NIOSH on May 10, 2010. Enhancing NIOSH's role early in the response and engaging NIOSH in any future revisions to federal response plans will ensure a timely and efficient health and safety response to such an event.

Because NIOSH is not organized or resourced to respond on a level of the magnitude required for DWH Response, assessment and planning on the future role of the agency in such large events is needed to align stakeholder expectations and NIOSH's capabilities.

Rostering Activities

Rostering is a mechanism to account for all workers engaged in a response. Integration of rostering into the functions of the incident command system (ICS) would be beneficial, especially if it were addressed as part of a national emergency/disaster preparedness process (including planning and training).

Rostering should begin as soon as possible to ensure that all workers have the opportunity to participate, and to avoid inefficiencies associated with identifying and contacting workers hired prior to the start of a rostering effort.

To facilitate rostering and data management, consideration should be given to creating financial vehicles that can be utilized just-in-time to conduct such efforts.

Health Hazard Evaluations (HHEs)

NIOSH posted interim HHE Reports on the website in a timely manner, which resulted in quick dissemination of data to health and safety representatives and the National Incident Command. However, the reports may have had limited utility in communicating with workers themselves; solutions to this challenge are discussed in the report.

Due to the urgency of deploying NIOSH staff, survey tools were developed quickly. Pre-development of general survey tools that can be modified during the event and sending senior field investigators to conduct initial qualitative risk evaluations may help improve the quality of the survey tool.

Technical Guidance and Communications

Having on-the-shelf documents and guidance that can be readily adapted and submitted for clearance can lessen the time necessary to make information available.

Standardization of industrial hygiene data collection, record keeping, and information posting among all entities conducting exposure assessments would enhance the value and potential use of exposure measurements, including linkage of exposure and potential health effects to specific job tasks.

Health Surveillance Activities

Existing public health surveillance systems and most disaster surveillance efforts are not designed to specifically address occupational health and safety outcomes, and adaptation of these systems at the time of the event is problematic for several reasons discussed in this report.

Maximum advantage should be taken of existing data streams, such as OSHA 300 logs, reports from state/local health departments, as well as logs from occupational health clinics, for injury and illness surveillance of response workers. Surveillance of these data streams should be coordinated within the ICS command structure as early in the response as possible.

Development of standardized instruments for baseline occupational surveillance and post-event occupational data collection could facilitate future response efforts to better recognize and understand injury and illness rates and allow corrective actions/interventions. A Worker Health Survey was developed during this event to assist in real-time surveillance efforts, but the response was ending before it could be implemented. However, this survey could be a tool adapted for future events.

While NIOSH has the expertise to develop tools and plans, insufficient in-house capacity within NIOSH is available to conduct large-scale surveillance efforts, and the existing contract mechanisms impose time constraints that impede immediate implementation of large-scale surveillance during disasters.

Other Health Assessment Efforts

Development of triggers or indicators in advance of a disaster will help make decisions as to whether long-term studies may be needed following a disaster.

Development of a decision matrix on the use of biological monitoring (collection of blood, urine, and other biological samples from workers) could help to rapidly determine whether biological monitoring would be beneficial for response workers.

Strategies and dissemination platforms for mental and behavioral health need to be further developed and incorporated into health and medical guidance for response and recovery operations.

NIOSH Response Activities

The following sections outline NIOSH's response activities following the DWH disaster. Occupational safety and health staff from NIOSH arrived on site in response to the oil spill on May 3, 2010, as part of the federal interagency effort (at the invitation of OSHA), to anticipate and address occupational safety and health needs in the Gulf Coast. The unprecedented circumstances and magnitude of the disaster posed numerous challenges for NIOSH and its federal, state, and local partners.

NIOSH staff deployed to multiple locations in the five-state Gulf region; coordinated cross-agency and interagency operations in Atlanta, GA, and Washington DC; supported other agencies' missions as subject matter experts; and provided technical and administrative support from NIOSH divisions, labs, and offices located throughout the country. This effort was a large activation of NIOSH personnel, cumulatively deploying 106 staff into the field and involving close to 250 staff in total. All NIOSH staff were medically cleared and received a pre-deployment briefing prior to deploying. NIOSH performed multiple activities to protect worker safety and health in the Gulf, as highlighted below.

NIOSH Rostering Effort

Thousands of workers engaged in the on- and off-shore containment and clean-up activities of the DWH response. NIOSH supported the UAC by establishing a roster of workers participating in response clean-up efforts, including volunteers. The roster information was collected at the time of training, or for those already trained, by working with on-site safety and supervisory officials to collect information from on-site workers at staging areas.

The concept for the worker roster was developed previously by an interagency work group coordinated by NIOSH. This group recommended that roster information be collected prospectively, as the World Trade Center experience highlighted the difficulties of obtaining roster information retrospectively. NIOSH developed this prospective roster to create a record of those who participated in the DWH response clean-up activities, collect information on the nature of work assignment and training received, and establish a mechanism to contact them about possible work-related symptoms of illness or injury, as needed. The UAC fully supported the rostering activity. Under the Paperwork Reduction Act, data collection instruments, including the roster, needed to be approved by the Office of Management and Budget (OMB). Through an emergency clearance mechanism, NIOSH was able to quickly obtain OMB clearance.

Workers could be rostered in one of three ways: 1) during safety training at official training sites (prior to being hired); 2) by NIOSH field staff visiting staging areas (locations to which trained workers reported for duty each day) in Louisiana, Mississippi, Alabama, and Florida; or 3) through a website. Individuals conducting the safety training sessions were provided with information about the NIOSH rostering effort (which they would convey to trainees), along with copies of the form and disclosure document to hand out to trainees during the class. All forms were available in English, Spanish and Vietnamese. The trainers collected the completed forms and mailed them back to NIOSH. Since rostering at training sites began after many workers had already been trained and assigned to a work location, NIOSH also deployed field teams to worksites and staging areas in the four affected states to attempt to roster workers at these locations. Additionally, NIOSH rostered oil spill response workers online through a website that had provisions to secure personal data. NIOSH provided a website link to multiple federal agencies, health departments, and BP and asked them to refer workers to the website to complete the rostering form electronically. This served as the mechanism for rostering staff who worked out of command posts located in Houma, LA (for the state of Louisiana) and Mobile, AL (for Alabama, Mississippi and Florida). In all cases respondents were informed that their information would be kept private to the extent authorized by law and that the information collected would be maintained in a secure manner. The initial paper or web-based rostering form took workers approximately 5 minutes to complete.

All completed roster forms were entered into a central database, recoded as necessary, and reviewed for errors. More than 55,000 workers completed the roster form. A final report detailing the demographics of the worker population rostered is complete and can be found at http://www.cdc.gov/niosh/docs/2011-175/pdfs/2011-175.pdf. NIOSH established a policy that allows qualified, external researchers to recruit individuals included in this roster for participation in potential future studies of possible persistent or long-term health effects.

Conducting Health Hazard Evaluations (HHEs)

On May 28, 2010, BP requested HHEs of DWH Response workers involved in the oil spill response, for both on-shore and off-shore activities. The NIOSH HHE program conducts investigations of potential workplace health problems in response to requests from workers, labor representatives, or employers. In response to these requests, the HHE team coordinated and conducted the analysis of hundreds of complex worker breathing zone air samples and other industrial hygiene samples. As information from the HHEs was analyzed, the findings were reported to BP (the HHE requesters) and the UAC, and were then posted to the NIOSH website. The ninth and last in a series of interim reports from this health hazard evaluation was issued in December 2010 and a final summary report was subsequently posted; all reports can be found on the following site: http://www.cdc.gov/niosh/topics/oilspillresponse/gulfspillhhe.html.

NIOSH made recommendations related to 1) work practices and personal protective equipment use to minimize potential for skin contact with oil residue; 2) heat stress management, including consideration of the role personal protective equipment may have in adding to workers' heat stress risk; 3) work practices and use of tools for beach cleaning; 4) adapting to unusual local circumstances that may affect implementation of occupational health management plans; 5) routine reporting of illnesses and injuries; 6) the need for pre-placement medical evaluations; and 7) work organization, stress, and mental health issues.

NIOSH Technical Guidance and Communications Activities for the Deepwater Horizon Response

During response efforts, it was critically important to develop and efficiently disseminate current and practical information for protecting response workers to enable management, workers, and health professionals to recognize and prevent acute illness and injury and mental distress as well as potential long-term physical and mental health effects. This information needed to be available for decision-makers, federal entities, state and local public health agencies, employers, workers, and labor representatives.

NIOSH devised new guidance and adapted existing documents to communicate potential risks associated with oil spill response work and protecting response workers, such as preplacement evaluations to avoid exacerbating underlying chronic illness (mental and physical) or injuries, managing the extremely hot and humid work environment, handling exposures to crude oil and dispersant chemicals, managing incident stress and fatigue, and devising sampling strategies for exposure monitoring. More than twenty interim guidance documents and communication products were developed and maintained on the NIOSH emergency response website in an expedited period. A NIOSH blog on the Deepwater Horizon Response was established to keep lines of communication open with workers and stakeholders.

NIOSH and OSHA jointly developed and issued a comprehensive health and safety guidance document (NIOSH/OSHA Interim Guidance for Protecting Deepwater Horizon Workers and Volunteers). The joint branding improved the visibility of the guidance and ensured consistency in federal messages. This guidance defined the administrative and engineering controls, as well as personal protective equipment, necessary to protect workers and volunteer responders.

NIOSH staff assessed the quality of the industrial hygiene and occupational health sampling data collected by other groups and issued comprehensive guidance on proper methods, sample collection and recordkeeping practice.

NIOSH staff also participated in regular UAC-BP-OSHA worker safety meetings to discuss current and emerging issues and provide recommendations in a timely manner.

Health Surveillance

NIOSH analyzed and prepared reports of UAC/BP Injury and Illness data; these can be found on the following website: http://www.cdc.gov/niosh/topics/oilspillresponse/data.html. The goal of this effort was to produce periodic overviews of the injuries and illnesses occurring among DWH responders. The reports were designed to provide a cumulative, broad-based overview of responder injuries and illnesses, identify injury patterns and trends of concern, and to inform a wide array of stakeholders interested in DWH responder safety and health, including UAC safety officials, federal partners such as OSHA and NIEHS, relevant state health departments, unions and other worker groups, and the general public.

Safety incident forms were completed by safety officials located throughout the Gulf area and forwarded to a central location. These forms were completed for any incident that affected the safety and health of a gulf responder, including federal responders, contractors, BP employees, and volunteers. NIOSH received a compilation of abstracted safety incident data on a weekly basis and used these data as the basis for its periodic reports. These data were reviewed, formatted, and coded by experienced Occupational Injury and Illness Classification System (OIICS) coders. Epidemiologists then used these coded data to produce a series of tables and graphs for inclusion in the report and analyzed the data for notable patterns and trends. These reports were posted on the NIOSH DWH Response Web page.

NIOSH Toxicity Testing Related to the Deepwater Horizon Response

During and after the DWH, in public meetings, press interviews, and other forums, many observers expressed a great deal of concern about the extent and potential effects of worker and community exposures to the crude oil, dispersants, and combination mixtures of the two. NIOSH initiated acute animal toxicity studies on the dispersant (Nalco Corexit 9500A), crude oil obtained from the source, and dispersant/crude oil mixtures. Inhalation studies measured pulmonary, cardiovascular, and central nervous system outcomes in laboratory test mice. Additionally, dermal exposure studies were conducted to assess hypersensitivity and immune-mediated responses in laboratory test mice. By conducting these animal toxicity studies, NIOSH hopes to contribute to the body of science on the potential health effects of exposures to crude oil, dispersant, and mixtures. Findings are published in conference abstracts and peer-reviewed journals. Links to this information can be found at http://www.cdc.gov/niosh/topics/oilspillresponse/chemDispersant.html

Lessons-Learned Observations

Lessons-learned in response to this disaster included those related to general deployment issues, rostering activities, HHEs, technical guidance and communications, health surveillance activities, toxicity testing, and other health assessment efforts. The collaborative work conducted with OSHA, HHS/ASPR, HHS/SAMHSA, HHS/NIH/NIEHS, the U.S. Coast Guard, state partners, and BP from the early stages of the response was key to efforts to protect response workers. NIOSH's overall approach to follow-up actions is in the final section of this report.

General Deployment Issues

NIOSH involvement in the DWH response and collaboration with the UAC began when NIOSH accompanied OSHA to a federal stakeholder meeting, at OSHA's invitation. Through this effort, NIOSH was able to sign an agreement with the Federal On-Scene Coordinator to roster workers, identify potential health effects associated with response work, and provide technical assistance to identify trends in injury and illness information. It is critical to have occupational safety and health expertise brought to the table early in an event and to have worker safety and health guidance prepared in advance. Any changes to federal response plans should include occupational safety and health experts in the planning process and a focus on fully integrating worker safety and health, including surveillance.

NIOSH is not organized or resourced for emergency response assistance on a level of the magnitude required for DWH Response without diverting time, personnel, and resources from its ongoing research programs and projects. Assessment and planning on the future role of NIOSH in such large-scale emergency responses is needed to align stakeholder expectations and NIOSH's capabilities. Internally organizing and deploying staff at this level and sustaining this effort for the duration of the response was challenging, albeit successful, in meeting the needs and demands for our services during this response. NIOSH deployed an unprecedented number of staff (106 employees) over a 3-month period, and nearly 20% of the Institute staff had some involvement in the response. Identifying and deploying staff that had suitable skill sets, met medical clearance criteria, and were available to deploy, was a large task that also disrupted on-going research and other activities. Recruiting qualified staff from other organizations to support occupational safety and health activities should be built into NIOSH's future response plans.

Rostering Activities

Rostering of response workers is an essential tool that could be used to contact workers for real-time health surveillance, depending on the length of the response, and for potential long-term follow-up of health status. NIOSH and others realized the value of rostering as a lesson learned from the World Trade Center emergency response. The rostering project was intended to be completed during mandatory worker training. However, many clean-up workers had already completed training before rostering was incorporated into training. This required a labor intensive effort by NIOSH staff to deploy across the Gulf Coast staging areas to conduct rostering. Lessons from the rostering effort include the following:

- Begin worker rostering immediately and integrate it into response activities as soon as possible to ensure that all workers have the opportunity to participate;
- Have a ready-to-use roster form prepared that can be quickly adapted and cleared; this will reduce the time necessary to initiate rostering;
- Direct the rostering program through the incident/unified command;
- Explore the feasibility of incorporating rostering into existing response programs (e.g., personnel accountability and training programs) to improve efficiency of the activity; and
- Develop mechanisms to encourage and facilitate employer participation.

Two key rostering issues included the number of paper rostering forms requiring manual input and the creation of an on-line system for field input. Initial estimates placed the number of expected rostering forms to be between 3,500 and 10,000 but in reality over 55,000 rostering forms were collected. A web-based roster input system was developed to provide an alternative means to complete rostering and to reduce data entry demands associated with the paper forms. Unfortunately, only 162 forms were collected via the web-based input system. This was a disappointing response based on the level of effort required for system design and implementation, but the materials may be preserved for future events. Barriers to utilization of the web-based form should be explored. Such a large volume of paper forms makes data entry and ultimately data utilization quite slow. Electronic data collection mechanisms such as hand-held devices, appropriate for field use, should be explored so that data can be collected and used in a more timely way.

To facilitate rostering and management of collected data, consideration should be given to creating financial vehicles that can be utilized just-in-time to conduct such efforts.

Conducting Health Hazard Evaluations (HHEs)

Interim HHE reports were posted on the website in a timely manner and distributed to a wide variety of agencies and entities identified as interested parties. The information was useful to health and safety professionals and policy makers, who likely had access to the on-line reports and who are often the necessary intermediaries in applying the recommendations from a health hazard evaluation in the field, and in communicating these essentially technical documents to a non-technical worker audience. However, two factors limited the utility of the reports for communicating information to workers themselves. Response workers in the field largely did not have internet access, and there was no other immediate means to disseminate the information to them, especially those in remote locations. Even if copies of the reports were more widely available, they may or may not have directly been useful to workers, especially those with limited time to read the documents or limited ability to adopt recommended interventions. Potential solutions identified in the field but not implemented include: 1) working through a local AM radio station that reportedly catered to fishermen to disseminate findings of the HHEs; and 2) developing large posters, in multiple languages and simplified terminology, for hanging in common areas (dining tent, recreation tent, etc.) to advertise/explain the HHE program goals, purpose, methods, etc. and to post information from interim reports. Additionally, it will be important to get these messages into safety briefings and safety plans as the response evolves. NIOSH should explore the development of a poster template that could be quickly tailored and used both in future responses and routine HHEs to improve communication with and access to workers.

NIOSH staff were quickly deployed to the field to get a snapshot epidemiological view of health symptoms and hazards. This required the development of survey materials for hazard assessment in a short time-frame. Developing survey tools in a hasty manner can inadvertently lead to not capturing important worker health information. Learning more about conditions in the field before survey development may help improve the quality of the survey tool. Pre-development of general survey tools and evaluation methods that can be modified during the event and deploying senior field investigators to conduct initial qualitative risk evaluations should be beneficial for future events. The survey tools developed for this event should be helpful for future events.

NIOSH Technical Guidance and Communications Activities for Deepwater Horizon Response

Utilization of the General Service Administration's federal website (www.data.gov/restorethegulf) was a productive and beneficial application of the federal partnership to provide public information. Federal entities engaged in future responses should consider using such a centralized tool for public access to information.

Various organizations posted their industrial hygiene sampling results, but the data were sometimes difficult to interpret. For example, sampling protocols, analytical methodologies, and/or explanatory text, such as job activity or work location relative to the presence of oil, were not always provided. Standardization of data collection and information posting would enhance the value and potential use of exposure measurements, including linkage of exposure and potential health effects.

To lessen the time for information to be made available, having on-the-shelf boilerplate documents, templates, and guidance that can be readily adapted to a specific situation is one step that can be taken to improving communications.

Health Surveillance Activities

Existing health surveillance systems have been established to help scientists and public health professionals track specific types of illnesses and injuries. These systems often are mutually exclusive of one another, may rely on different definitions and information technologies, and generally are designed to monitor well-recognized or acute conditions, such as infectious diseases or cases of poisoning. These different systems usually perform well for their respective purposes. However, data from one system may not be compatible with data from another, subtle or long-term effects may not be detected, and most disaster surveillance systems, including syndromic surveillance systems, are not designed to address occupational health and safety outcomes including fatalities, traumatic injury, heat stress, psychological and behavioral correlates of distress, and musculoskeletal injuries and other adverse exposure events. Existing surveillance systems may provide, at best, an incomplete picture of responders' injuries and illnesses in a large-scale emergency such as the DWH Response. Adaptation of established surveillance systems at the time of the event to capture event-specific data is problematic as it is difficult to implement quickly enough to be useful and will lack baseline or comparison data unless the adapted system is also used in unaffected areas. Continued attention needs to be placed on developing surveillance systems that can better track occupational health outcomes during emergent events.

Maximum advantage should be taken of existing data streams that could be used for health surveillance of response workers during the response. Data may be obtained from monitoring the OSHA 300 logs of responding agencies, reports from state and local health departments, as well at the logs of occupational health clinics, first aid stations, etc. Dispatch records and ambulance run records of EMS agencies providing service to the response worksites could be monitored as well. Surveillance of these data streams should be coordinated within the ICS command structure as early in the response as possible.

As a part of emergency preparedness planning, federal, state, and local agencies should give consideration to the development of standardized instruments for baseline occupational surveillance and post-event occupational data collection and analysis that could be easily adapted to specific events and used by various organizations. Once a response has been initiated, provisions should be made to capture the more unusual or dynamic components of a response and its aftermath.

Improved occupational injury and illness surveillance may be achieved through enhanced integration and coordination with other surveillance activities at the HHS/Centers for Disease Control and Prevention (CDC) and other agencies within HHS.

Other Health Assessment Efforts

Biological Monitoring. In many emergency responses, including the one for DWH, questions arise about the need for biological monitoring (collection of blood and urine for testing of exposure to chemicals). Determining when biological monitoring should be conducted can be difficult in part because it may not be clear as to whether a scientific rationale exists for biological monitoring in a given situation, or whether results from the monitoring can be interpreted meaningfully or reliably. Additionally, a fundamental tenet related to a decision to recommend biological monitoring is that it should provide a clear benefit to the worker. Formulation of a rationale and purpose for biological monitoring therefore needs to be rapidly determined early in the event. Although NIOSH staff held ongoing internal discussions on biological monitoring during the DWH response, it would have been helpful to have had a pre-developed decision matrix to help answer questions about 1) the purpose of biomonitoring (how the results will be used); 2) likelihood and impact of dermal exposures (that are not easily assessed by traditional industrial hygiene methods); 3) efficacy of personal protective equipment; 4) health risk associated with exposure(s); 5) the future consideration of health outcomes; and 6) the existence of feasible biomarkers. In conjunction with developing this matrix for worker populations, other groups may want to consider a similar approach for non-occupational settings.

Long-Term Studies. In many emergency responses, including the oil spill response, NIOSH and others are frequently faced with the question of whether long-term health studies of workers should be conducted. Long-term studies of workers following an event can be costly and difficult to design; therefore, one must consider whether the study will produce useful, reliable results. The need for long-term studies should be assessed in the beginning of the event and should be periodically revisited during an event as worker job activities, safety hazards, exposures and response events will change during the course of the event. To facilitate decision making about the need for long-term studies, it would be useful for scientists and others to identify beforehand the factors that would prompt such an effort in a given situation. Factors or triggers for initiating a long-term study could include the following:

- Documented exposures to hazardous agents among workers
- Concerns that there may be long-term health effects among workers
- Uniqueness and magnitude of the event
- Pre-identified areas of research interest
- Unique vulnerability of worker population
- Toxicological properties of the materials involved

Initiating a study is not necessarily based on the number of workers involved in the response or a lack of literature on health effects of exposure. The triggers may vary depending on the event. Triggers may also vary depending on the purpose of the long-term study. For example, a study may be conducted for the purpose of medical monitoring, surveillance, or characterizing possible health effects of a novel exposure. Pre-identifying areas of research interest will provide a more thoughtful and deliberate agenda for study. NIOSH is leading an interagency workgroup that includes development of event triggers for conducting long-term studies. The development of such triggers should help the decision making for appropriate long-term studies.

Mental Health. Mental and behavioral health activities are addressed across multiple HHS agencies. Worker health issues need to be prominently incorporated with a common framework of HHS operations that addresses behavioral health of affected populations in an emergency response operation. Continued development of strategies and dissemination platforms are needed to incorporate these issues in health and medical guidance and response/recovery.

Follow-up Actions and Next Steps

This report, while identifying numerous lessons-learned and some possible remedial approaches, does not specifically outline an overall action plan because resolution of several of the lessons-learned will require significant resources and coordination with other government entities. However, NIOSH is developing an action plan to address those issues that are within its control and will not require extensive resources to accomplish.

Additionally, NIOSH sponsored an interagency work group that, over the past several years, developed a guidance document to address responder health monitoring and surveillance. The guidance provides a comprehensive set of strategies for enhancing the safety and health of responders to help managers, medical personnel, and health and safety representatives prepare thoroughly before an event and subsequently help ensure worker health and safety during and following an event. The document contains two main sections: 1) a Guidance section that includes guidance and recommendations during the pre-, during-, and post-stages of deployment; and 2) a Tools section that provides links to relevant existing documents and examples of materials that could be used in a response (e.g., surveys and standardized questionnaires, checklists, databases, and software programs). Among the various areas addressed are: Health Screening, Rostering, Training, Credentialing, Exposure Assessment and Controls, Medical Monitoring, and Medical Surveillance. The overall objective of the guidance is to help ensure only medically cleared, trained, and properly equipped personnel are selected for deployment, their work environment and health is effectively monitored and surveilled throughout the event, and provisions are made for post-event health medical monitoring and surveillance where indicated. A public draft of this document was published in the Federal Register for public comment and has been updated based on these comments. The document has been submitted to the National Response Team for review and is currently available at http://www.cdc.gov/niosh/docket/review/docket223/

Acknowledgements

The following NIOSH staff, listed alphabetically, provided significant contributions and/or authored sections of this report:

Bruce Bernard, Medical Section Chief, Hazard Evaluations and Technical Assistance Branch (HETAB), Division of Surveillance, Hazard Evaluations, and Field Studies (DSHEFS)

Vincent Castranova, Branch Chief, Pathology and Physiology Research Branch, Health Effects Laboratory Division

Gayle DeBord, Associate Director for Science, Division of Applied Research and Technology

John Decker, Senior Scientist, Office of the Director

Lisa Delaney, Deputy Associate Director, Emergency Preparedness and Response Office (EPRO)

Renee Funk, Veterinary Epidemiologist, EPRO

John Gibbins, Veterinary Epidemiologist, HETAB, DSHEFS

Bradley King, Industrial Hygienist, HETAB, DSHEFS

Margaret Kitt, Deputy Director for Program

Dori Reissman, Senior Medical Advisor, World Trade Center Program, Office of the Director

Teresa Seitz, Supervisory Industrial Hygiene Team Lead, HETAB, DSHEFS

James Spahr, Associate Director, EPRO

Marie Haring Sweeney, Branch Chief, Surveillance Branch, DSHEFS

Allison Tepper, Branch Chief, HETAB, DSHEFS

www.ingramcontent.com/pod-product-compliance
Lightning Source LLC
Chambersburg PA
CBHW081822170526
45167CB00008B/3501